CHINA

Shanghai

Pacific
Ocean

Sichuan
Province

Yangtze
River

Indian
Ocean

A LITTLE ROUND PANDA ON THE BIG BLUE EARTH

To Bryan, although we didn't see pandas, China
will always remind me of you—T.C.

To Michael, Alexander, and Nicolas, always.
To Emilia: bem vinda!—L.N.P.

Library of Congress Cataloging-in-Publication Data

Names: Christie, Tory, author. | Powell, Luciana Navarro, illustrator.

Title: A little round panda on the big blue earth / by Tory Christie ; illustrated by Luciana Navarro Powell.

Description: First edition. | Mankato, MN : Amicus Ink, 2021. | Audience: Ages 4-8 | Summary: "Starting with a baby panda in a bamboo forest, this richly illustrated poem illuminates a unique geography perspective, showcasing China's environment with ever-widening views from forest to terraced farms to village and city, country, continent, ocean, and finally the planet in space. Endsheets include a map of Asia labeling places shown"-- Provided by publisher.

Identifiers: LCCN 2019046018 (print) | LCCN 2019046019 (ebook) | ISBN 9781681526546 (hardcover) | ISBN 9781681526553 (pdf)

Subjects: LCSH: Pandas--Juvenile literature.

Classification: LCC QL737.C214 C47 2021 (print) | LCC QL737.C214 (ebook) | DDC 599.789--dc23

LC record available at https://lccn.loc.gov/2019046018

LC ebook record available at https://lccn.loc.gov/2019046019

Editor: Rebecca Glaser | Designer: Aubrey Harper

First edition 9 8 7 6 5 4 3 2 1

Printed in China

A LITTLE ROUND PANDA ON THE BIG BLUE EARTH

by Tory Christie

illustrated by Luciana Navarro Powell

amicus ink

Mankato, Minnesota

A little
round
panda

Munches
on bamboo

Sprouting
from the
damp
ground

On a hill
covered
in mist

Trickling
onto a
muddy path

That twists

and turns

and wanders

To a murky river

Where a family
climbs into a boat

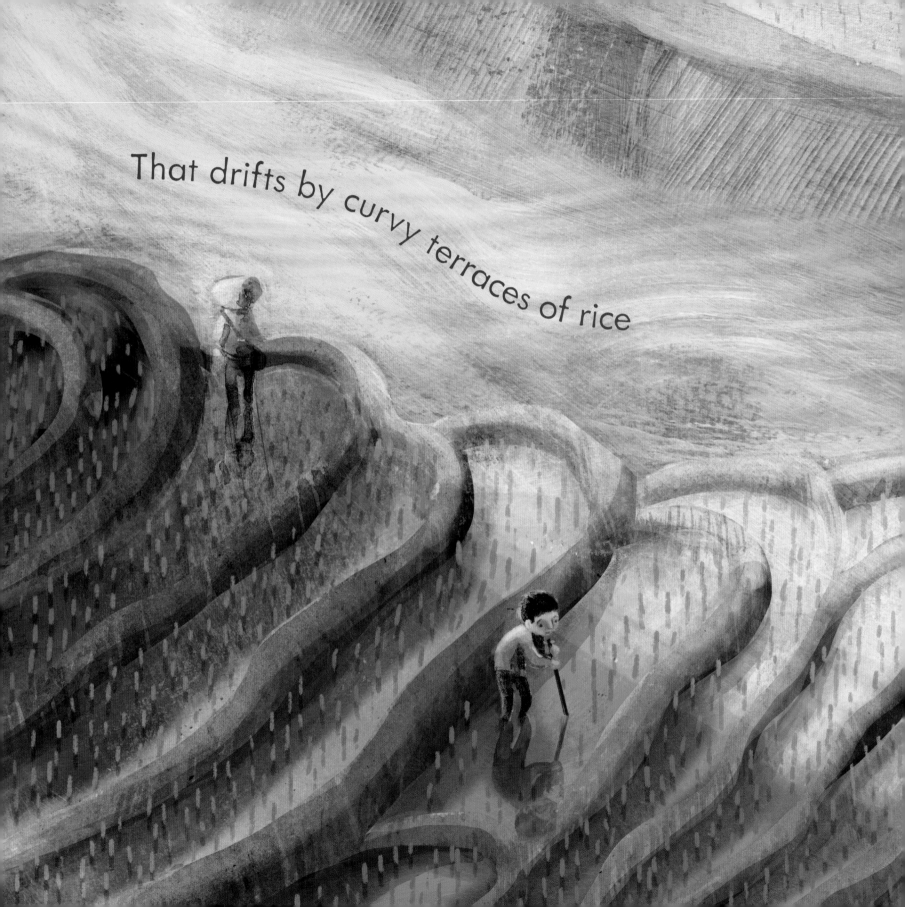

That drifts by curvy terraces of rice

And farms
and villages

Scattered across
an ancient country

All the way to a bustling city

of a great continent

With lights
blink-blink-blinking

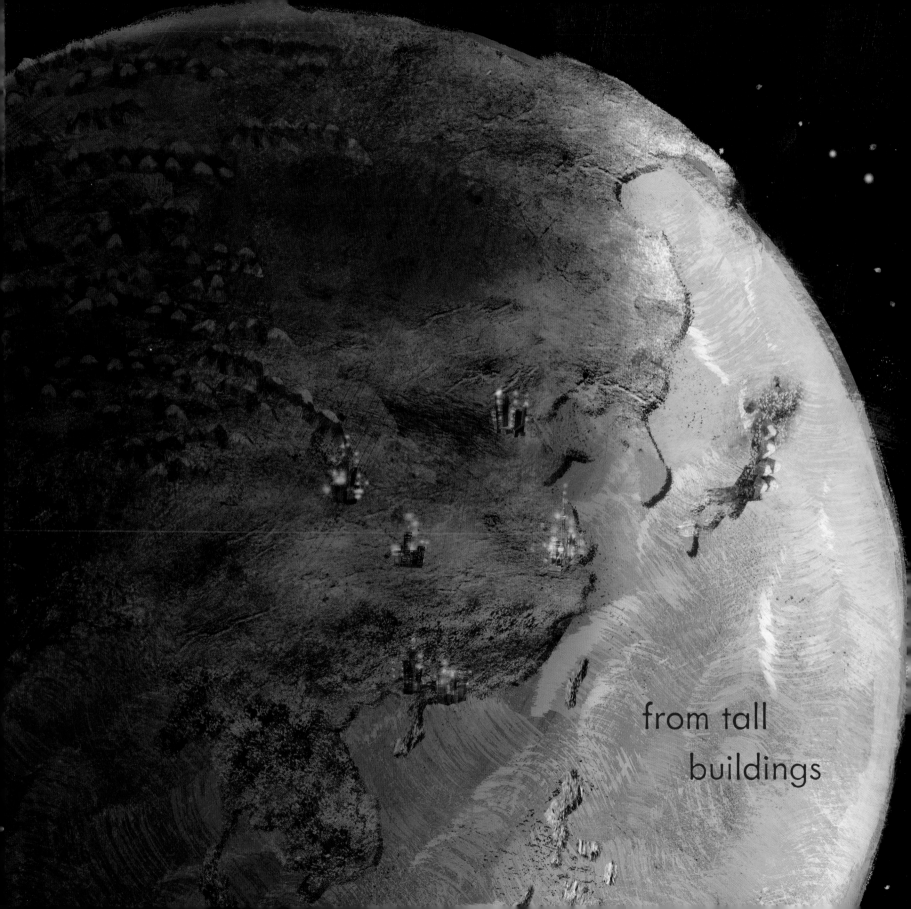

from tall
buildings

On the **big** blue Earth . . .

...where a little round panda munches on bamboo.

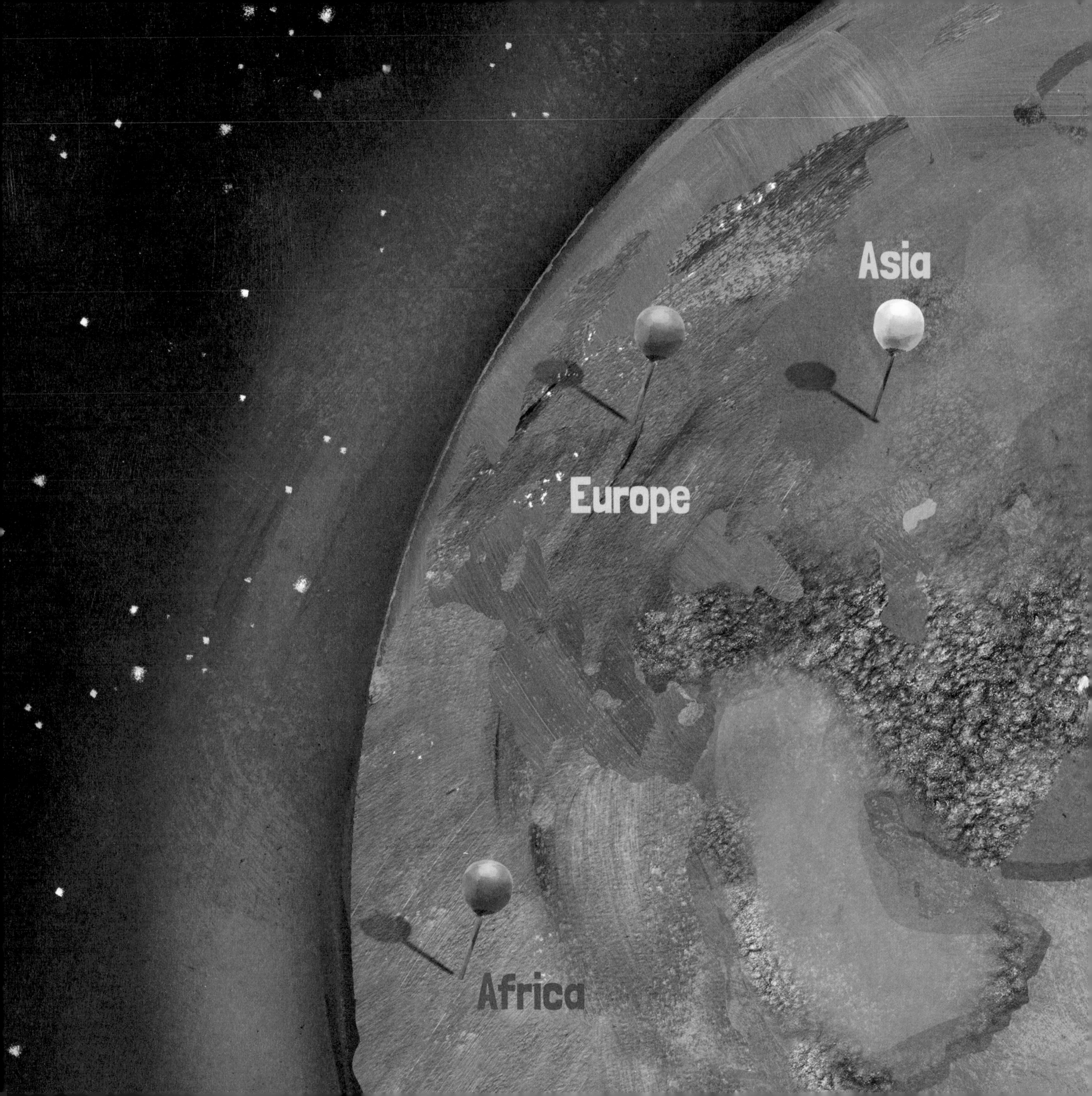
Asia
Europe
Africa